Photo Field Guide to the

Freshwater Mussels

of Ontario

Dedicated to the memory of Ian Carmichael

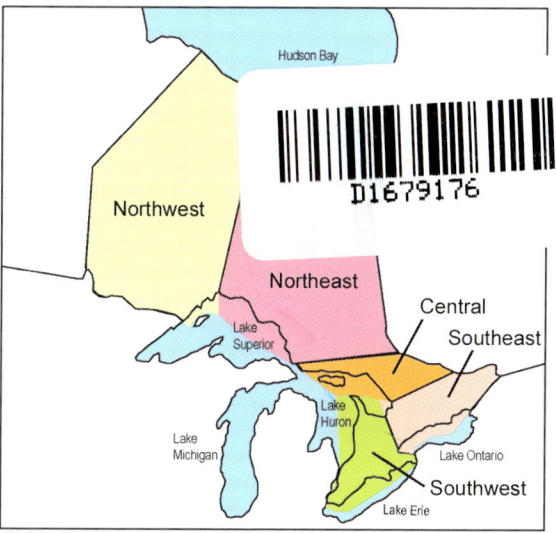

Janice Metcalfe-Smith

Alistair MacKenzie

Ian Carmichael

Daryl McGoldrick

Copyright © 2005
Janice Metcalfe-Smith, Alistair MacKenzie, Ian Carmichael and Daryl McGoldrick.

All rights reserved. No part of this book may be reproduced in any form or by any means, electronic or mechanical, including photocopying, recording, or by any information storage and retrieval system, without written permission of the publisher.

First published in 2005 by
> St. Thomas Field Naturalist Club Incorporated
> Box 23009
> St. Thomas, ON, Canada
> N5R 6A3

Cover Photographs:

Background - Sydenham River near the village of Alvinston, Lambton County.
Mussel species - Top: Deertoe - *Truncilla truncata*
Middle: Mapleleaf - *Quadrula quadrula*
Bottom: Purple Wartyback - *Cyclonaias tuberculata*

ISBN 0-9733179-2-2

Printed in Canada by Aylmer Express
Reprinted 2010, 2012, 2013, 2015, 2017, 2019

Preface

Freshwater mussels are a fascinating group of invertebrates that live on the bottom of streams, rivers, lakes and ponds. They are not readily seen by the casual observer, but professional biologists and naturalists know that these "living filters" play an important role in aquatic ecosystems by cleaning the water and providing food for a variety of fishes and wildlife. Mussels have a remarkable life cycle in that they are parasitic on a vertebrate host, typically a fish, during the larval stage. Unfortunately, freshwater mussels are among the most endangered organisms in North America. Mussels are threatened by many human influences including pollution, habitat destruction and the introduction of aquatic invasive species. In Ontario, 28 of our 41 native species are showing signs of decline and many are in urgent need of protection and recovery efforts. This field guide is intended as a tool for professional biologists, naturalists, students and members of the general public who are interested in learning about freshwater mussels and how to identify the species that are native to Ontario. By fostering an understanding and appreciation of mussels and their role in aquatic ecosystems, we hope to gain support for conservation programs designed to preserve this important part of our natural heritage.

A live Flutedshell (*Lasmigona costata*) filter-feeding on the bottom of the North Thames River. The posterior end of the mussel is exposed showing the exhalent (right) and inhalent siphons.

Acknowledgments

We would like to thank all who helped make this guide possible, especially Philip McColl of the Graphic Arts Section, National Water Research Institute, for drafting the illustrations and creating the digital scans of mussel shells using a 3-D object scanning system. Thanks to Don Stacey, Department of Natural History (Invertebrates), Royal Ontario Museum, for lending us specimens of the Cylindrical Papershell, Eastern Floater, Hickorynut, Snuffbox and Triangle Floater to be scanned for the guide and also to Dr. Jean-Marc Gagnon, Chief Collection Manager, Invertebrate Collections, Canadian Museum of Nature, for lending us several specimens of the Elephantear. Unless otherwise noted, the photographs used in the preparation of this guide are from the collection of the National Water Research Institute and were taken by Shawn Staton, Daryl McGoldrick, David Zanatta or Janice Metcalfe-Smith. We are grateful to Muriel Andreae of the St. Clair Region Conservation Authority and Donald Sutherland of Ontario's Natural Heritage Information Centre for helping us obtain the funds to publish the guide. Special thanks to Environment Canada's Habitat Stewardship Program for Species at Risk, the National Water Research Institute, the Ontario Ministry of Natural Resources (Peterborough), the St. Thomas Field Naturalists and the St. Clair Region Conservation Foundation for funding the project. The following individuals provided helpful comments on the guide prior to publication and we greatly appreciate their input: Muriel Andreae, Tom Chatterton, Larry Cornelis, Becky Cudmore, Alan Dextrase, Dr. Jean-Marc Gagnon, Erin James, Dr. Gerald Mackie, Dr. André Martel, Dr. Todd Morris, Barry Myler, Michael Oldham, Don Stacey, Donald Sutherland, Kara Vlasman, Winifred Wake and David Zanatta.

 Environment Canada / Environnement Canada

 St. Clair Region Conservation Foundation

 St. Clair Region Conservation Authority

 Natural Heritage Information Centre

 The St. Thomas Field Naturalist Club Inc.

General Information

Classification

Freshwater mussels are molluscs, that is, members of the Phylum Mollusca. This large and diverse group of organisms also includes snails, slugs, clams, scallops, oysters, squids and octopuses. Molluscs are soft-bodied, non-segmented invertebrates that have a muscular foot for burrowing or crawling and an enveloping sheath of tissue known as the mantle that, in most species, secretes a calcareous shell. Freshwater mussels belong to the Class Bivalvia (sometimes known as the Pelecypoda, which means "hatchet foot"). Bivalves are a large group of freshwater and marine molluscs characterized by having a pair of hinged shells. Two types of freshwater bivalves occur in North America: the fingernail and pea clams, which belong to the Order Veneroida, and the freshwater mussels or "pearly mussels", which are members of the Order Unionoida. Although unionoids are technically mussels, many people refer to them as clams. North America supports the greatest variety of freshwater mussels on the planet – nearly 300 of the world's 1000 species. All North American species belong to either the Family Unionidae or the Family Margaritiferidae. There are 55 species of freshwater mussels in Canada and 41 of these occur in Ontario. All Ontario species are members of the Family Unionidae.

Internal Structures of a Freshwater Mussel

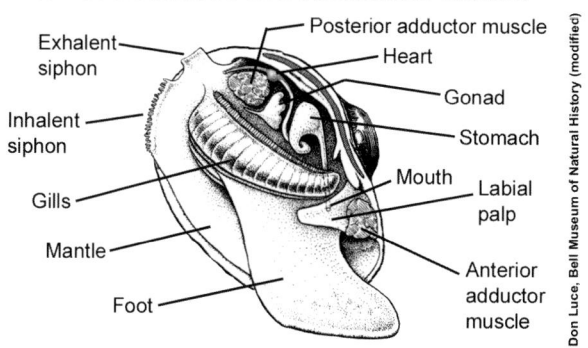

Two other types of freshwater bivalves have "invaded" North America. The Asiatic Clam (*Corbicula fluminea*) was intentionally introduced into Washington and California from southeast Asia in the 1930s as food for humans and domestic fowl. It has since spread throughout most of the United States, but has not done well in the Great Lakes because it cannot tolerate the cold water temperatures. The Zebra Mussel (*Dreissena polymorpha*) and its relative, the Quagga Mussel (*D. bugensis*), are native to the Caspian Sea. They were inadvertently transported across the Atlantic in the ballast water of an ocean-going ship sometime in the mid-1980s, then released into Lake St. Clair. They multiplied and spread rapidly and are now found

General Information (continued)

throughout much of eastern North America. Zebra Mussels pose a significant threat to native freshwater mussels (see page 9).

Biology and Life Cycle

Freshwater mussels are the largest and longest-lived freshwater invertebrates in North America. Many species reach lengths of 10 or even 20 cm and have life spans up to many decades. A rough estimate of a mussel's age can be made by counting the growth rings on its shell. Mussels occupy a wide range of permanent aquatic habitats including streams, rivers, ponds and lakes. They reach their greatest diversity and abundance in large rivers, which offer a constant supply of oxygen and food and a variety of habitat types. Mussels are essentially sedentary as adults, spending their lives partially or almost completely buried in the substrate. They feed by drawing in water through the inhalent siphon and passing it across the gills to filter out small particles such as algae, bacteria and detritus. These particles are delivered to two pairs of labial palps that sort out the food items for ingestion. Filtered water and waste are expelled through the exhalent siphon.

Reproductive Cycle of a Freshwater Mussel

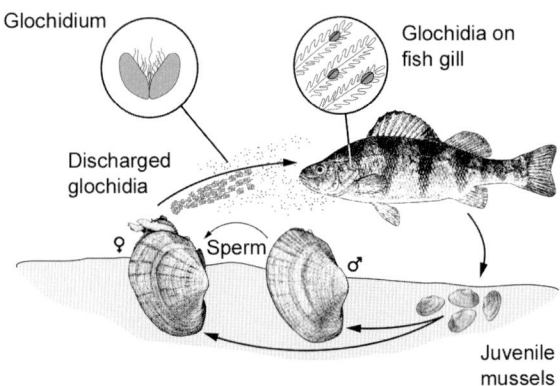

Freshwater mussels have a unique life cycle in that they are parasitic on fishes (or an amphibian in the case of the Salamander Mussel) during their larval stage. Most species have separate sexes although a few are known to be hermaphroditic, i.e., able to produce both sperm and eggs. During spawning, males release sperm into the water through their exhalent siphons, and females living downstream take in the sperm through their inhalent siphons. Eggs are fertilized in specialized portions of the female's gills called marsupia. The embryos remain in the gills until they reach an intermediate larval stage called the

General Information (continued)

glochidium. The marsupia become progressively more swollen and pad-like as the glochidia develop. The length of time required for larvae to reach this stage varies from species to species. In long-term brooders, spawning occurs in the summer or early fall and the glochidia are held until the following spring. In short-term brooders, spawning occurs in the spring and the glochidia are released later that same summer.

When conditions are right (temperature, photoperiod, time of year when the host is likely to be present), the female mussel releases her glochidia into the water where they must quickly attach to the gills or fins of an appropriate host. Specific hosts for many species of mussels in Canada are unknown. The Ohio State University's Mussel/Host Database (http://128.146.250.63/Musselhost/) provides a list of hosts for mussels in the U.S. Glochidia look like miniature mussels with a bivalved shell and a single adductor muscle. They become encysted in the tissues of the host and derive nourishment from its body fluids for a period of time ranging from 6 days to over 6 months, depending on the species. The glochidia transform into juvenile mussels during this parasitic phase. Once metamorphosis is complete, the juvenile ruptures the cyst and falls to the river or lake bottom to begin life as a free-living mussel. The juveniles, which are initially less than 0.5 mm in size, burrow deeply into the substrate upon release and remain there for the first few years of life. Juveniles of many species, as well as adults of some of the smaller species, produce mucous threads that become attached to pebbles or other hard objects and help prevent them from being swept away by water currents. Most mussel species have only a few suitable hosts and the likelihood of a single glochidium surviving to

A female Wavyrayed Lampmussel (*Lampsilis fasciola*) displaying a minnow-shaped lure - complete with eye spots and tail fin - to attract a potential fish host. The mussel is almost completely buried with only the edge of the shell and the lure visible.

General Information (continued)

maturity is extremely low. To improve on these odds, mussels produce millions of glochidia and some species have evolved specialized structures to attract their hosts. In females of the genus *Lampsilis*, for example, the mantle flaps often take the form of a minnow-shaped "lure". Members of the genus *Ptychobranchus* produce conglutinates, which are specialized packets of glochidia bound in a mucous matrix that mimic food items such as fish fry or insect larvae. Female Rainbows (*Villosa iris*) have a spectacular lure that imitates a crawling crayfish. Images of the glochidia and lures of a variety of species can be seen on the web site of the Unio Gallery at Southwest Missouri State University (http://courses.smsu.edu/mcb095f/gallery/).

Ecological Role and Human Uses

Streams, rivers and lakes in their natural unpolluted state may be literally paved with freshwater mussels. The filtering capacity of such dense mussel "beds" makes them natural water purifiers. Mussels remove large quantities of suspended particles such as algae, bacteria and other organic material from the water column and convert the nutrients into forms that are readily used by other aquatic organisms. Feces and pseudofeces (material filtered but not ingested) become a nutritious food source for other benthic organisms, while nutrients excreted in dissolved form provide food for plankton and aquatic plants. Mussels are eaten by a variety of mammals such as River Otters, Mink, Raccoons and especially Muskrats. Some birds and several fishes, including Freshwater Drum, sturgeons, catfishes and suckers, also consume mussels. As mussels move about in the substrate, they mix and oxygenate the sediment and stimulate microbial activity in much the same way as earthworms in a garden. Empty mussel shells serve as egg-laying sites for fishes and other animals, hiding places for small fishes and crayfishes, and attachment sites for algae and insect larvae. Freshwater mussels are useful early warning indicators of environmental degradation since they are sensitive to many kinds of pollution and habitat alteration. Because they are long-lived and accumulate many toxic substances, they can be used as biomonitors of environmental contamination.

Native North Americans harvested freshwater mussels for food, jewellery, utensils, tools and tempering pottery. Natural freshwater pearls were discovered in the mid-1800s and pearl hunting soon became popular. Beginning in the late 1800s, mussels were harvested in enormous numbers for their shells, which were used to make pearl buttons for clothing. Most of this activity was concentrated in the upper Mississippi River basin, but a small mussel fishery was established in the early 1900s in the lower Grand and Thames rivers in Ontario. The pearl button industry collapsed with the invention and widespread use of plastics in the 1940s. In the 1950s, the Japanese discovered that beads cut from the shells of some North American mussel species

General Information (continued)

and inserted into living oysters produced high quality cultured pearls. Tonnes of mussel shells are exported annually from the United States for the cultured pearl industry.

Conservation Status

Freshwater mussels are among the most endangered organisms in North America, with 65% of species at some risk of extinction and another 7% (21 species) already extinct. The general status of Canada's freshwater mussels was assessed in 2004. Results showed that 65% of our species are in need of conservation action – more than any other group of animals or plants assessed to date (for more information, visit the Wild Species web site at www.wildspecies.ca). Mussels are threatened by habitat loss and degradation due to dams, dredging, sedimentation and pollution, by the loss of their host fishes, and by climate change and aquatic invasive species, especially the Zebra Mussel. Zebra Mussels attach to the shells of native mussels by the hundreds or thousands, causing them to die from lack of oxygen or food. As a result, native mussels have been nearly eliminated from much of the Great Lakes system, including Lake St. Clair, Lake Erie, Lake Ontario and the Detroit, Niagara, Trent-Severn, Rideau and St. Lawrence rivers. Zebra Mussels could easily become established in many inland lakes and reservoirs if we do not take measures to prevent their transfer from Great Lakes waters. In fact, Zebra Mussels were discovered in two reservoirs on the Thames River in 2003.

A Plain Pocketbook (*Lampsilis cardium*) found in Lake St. Clair with a heavy infestation of Zebra Mussels (*Dreissena polymorpha*).

General Information (continued)

Ontario's Natural Heritage Information Centre (NHIC) in Peterborough, Ontario, compiles, maintains and provides information on rare, threatened and endangered species, vegetation communities and natural areas in Ontario. This information is used for ecologically sound land use planning and to support conservation activities. The NHIC assigns provincial (subnational) ranks (SRANKS) to species in Ontario based on the known or expected number of extant occurrences, the extent of range, population size and the degree of threat in the province. These ranks are not legal designations; rather, they provide an assessment of the relative conservation concern for species in the province. Species ranked S1 (Extremely rare), S2 (Very rare), S3 (Rare to uncommon), SH (Historically known) or SU (Unrankable due to insufficient information) are actively tracked and occurrence information is compiled for each. Species ranked S4 (Common) or S5 (Very common) are not tracked. Further information on the process of developing ranks and the status of species in other jurisdictions, and globally, may be found on the NatureServe web site www.natureserve.org/explorer/. The NHIC welcomes information on occurrences of actively tracked species. To report such information, use the Rare Species Report Form available at: www.mnr.gov.on.ca/MNR/nhic.

The Committee on the Status of Endangered Wildlife in Canada (COSEWIC) began to consider freshwater mussels and other molluscs for listing in 1995. To date, 8 of Ontario's 41 species have been designated as Endangered and many others are on the priority candidate list awaiting assessment. For more information about COSEWIC and the listing process, visit the web site at www.cosewic.gc.ca. Canada's *Species at Risk Act* (SARA) was proclaimed in 2003 and came into full effect in June, 2004. The purpose of SARA is to prevent wildlife species from becoming extinct and to provide for the recovery of species that are at risk as a result of human activities. Eleven species of freshwater mussels are among the 349 species on the List of Wildlife Species at Risk and are protected under the new law. More mussel species will likely be added to the list in the future. For more information on SARA, visit the *Species at Risk Act* Public Registry at www.sararegistry.gc.ca. Seven species of mussels appear on the Species at Risk in Ontario (SARO) list, but are not regulated because aquatic species fall under federal jurisdiction. All endangered and threatened species on the SARO list, whether regulated (categories END-R and THR-R) or not (categories END and THR), are afforded habitat protection under the Provincial Policy Statement of the *Planning Act* and under the *Aggregate*

General Information (continued)

Resources Act. The Ontario Ministry of Natural Resources (OMNR) is responsible for assigning status designations to native Ontario species. Current designations can be found on the OMNR web site www.mnr.gov.on.ca or www.ontarioparks.com/english/sar.html.

Mussel Collecting and the Law

Freshwater mussels are considered to be fishes (shellfishes) and are protected under the Ontario Fishery Regulations made under the federal *Fisheries Act*. These regulations prohibit the collection of live mussels of any species in Ontario without a permit from the Ontario Ministry of Natural Resources. It is an offence to undertake any activity that affects an aquatic species listed as Extirpated, Endangered or Threatened under SARA, without a permit from Fisheries and Oceans Canada. In the case of freshwater mussels, collecting and possessing even the shells of one of these species requires a permit.

Using an underwater viewer to search the riverbed for mussels.

How to Search for Mussels

Even if you have a permit, live mussels should not be sacrificed unless absolutely necessary. It is best to study them in the field and return them unharmed to the place where you found them. An easy way to determine if mussels are present is to walk along the lakeshore or riverbank during periods of low water and look for shells. Shells are easier than live animals to identify to species because you can examine both the internal and external shell features. Muskrats and other predators leave piles of shells called "middens" along the bank, in shallow water, or in sheltered areas such as under downed trees or near bridge abutments, so look for these as well. Once you have established that mussels may be present, you can use one of several

General Information (continued)

methods to collect them. In clear, shallow water they can be collected by hand-picking. Chest or hip waders, polarized sunglasses and a glass-bottomed bucket or underwater viewer, such as the one shown on page 11, are useful aids. Usually the only part of the mussel visible is the posterior end of the shell and the siphons. If mussels have been moving, they may leave a meandering trail in the sediment. In flowing water, always walk upstream so that the substrate you have disturbed will be flushed behind you and not obstruct your view. Snorkelling can be used in slightly deeper water provided it is clear. This method tends to be more effective for detecting very small mussels, and there is less chance of disturbing the substrate or trampling the animals. SCUBA diving is the best method to use in deep water. Be mindful of boat traffic, dangerous currents, poor visibility, cold water and other hazards when snorkelling or diving. Wading can also be hazardous in rivers that have variable water depths, unstable substrates or strong currents. Grabs and dredges are seldom used, as they are inefficient and destructive. If the site is wadeable but the water is murky, you may have to feel for mussels with your hands. Wear rubber gloves to protect your hands from broken glass, sharp pieces of metal and rusty fishing hooks.

Biologists searching a site on the East Sydenham River for mussels using a series of submerged $1m^2$ quadrats set in a grid pattern. This technique allows them to calculate the size and density of the mussel population.

General Information (continued)

Mussels should be kept in the water at all times to minimize stress. A mesh bag can be used to hold the mussels as you continue your search. Mussels should be handled gently and returned to the substrate as quickly as possible after collection. Place each animal back in the location where it was found, or if this is not possible then at least in an area where other live mussels are found in the same waterbody. Orient the animal with the posterior end up so it can quickly rebury itself and resume filtering. Avoid placing small or thin-shelled mussels in a spot where they could be swept away by currents. If you have difficulty identifying live mussels to species, take photographs from several angles (as illustrated in this guide) and show them to an expert. Mussel shells are often covered in algae or calcium carbonate deposits that can obscure the key features used in identification. A small, soft brush such as a nail brush is useful for removing this material. You can also collect empty shells and take them back with you for further examination. The Ausable-Bayfield, Grand River, St. Clair Region and Upper Thames River Conservation Authorities have collections of identified mussel shells from their respective regions that may be available for your reference. Remember that you need a permit to collect the shells of any of the SARA-listed species. For detailed information on sampling designs and methods for studying freshwater mussel populations, see Strayer and Smith (2003).

Guide Information

Ontario's 41 freshwater mussel species belong to three subfamilies of the Family Unionidae: the Ambleminae (9 species), Anodontinae (12 species) and Lampsilinae (20 species). The Ambleminae are characterized by thick, dark-coloured shells with few rays, heavy hinge teeth and little or no difference in shell shape between males and females. The Anodontinae have thin, generally greenish-coloured shells with faint or no rays, poorly developed or absent hinge teeth and no differences in shape between the sexes. The Lampsilinae have thin to moderately thick shells that tend to be yellowish in colour and frequently have bright green rays, well-developed (if sometimes delicate) hinge teeth and, in some species, distinct differences in shell shape between the sexes.

In the species accounts to follow, species are arranged by subfamily first, then alphabetically by scientific name within each subfamily. Scientific and common names follow Turgeon *et al.* (1998). Typical and maximum shell lengths for each species are shown in the upper right corner of each page. The lengths are largely based on over 20,000 live specimens collected and released in southern Ontario by the senior author and her study team between 1996 and 2003. Shell length refers to the maximum distance between the anterior and posterior ends of the shell. Mussels can be identified to species using external and internal features of the shell (see page 16). Shell shape and width are also important characteristics. Seven basic shell shapes and two basic widths are shown below, and these terms are used throughout the guide.

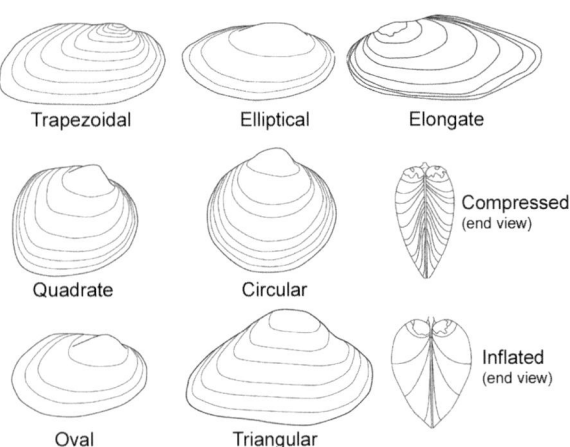

Trapezoidal   Elliptical   Elongate

Quadrate   Circular   Compressed (end view)

Oval   Triangular   Inflated (end view)

Guide Information (continued)

Scanned images of the exterior of the right valve and interior of the left valve are presented for each species. For species that are sexually dimorphic (males and females have a different shell shape), external and internal views of a male specimen and the external view of a female specimen are shown. For species that have subtle differences in shell shape between the sexes, only one external and one internal view are shown. Additional scans and photographs of live specimens are used to illustrate the range of shapes and colours that may be encountered, to show an example of a juvenile specimen if it looks substantially different from the adult, or to better illustrate a key feature. Close-up views of the beak sculpture or hinge teeth (pseudocardinal and lateral teeth) are included where these features are diagnostic. Three or more features that we find to be most useful for identifying each species are indicated by red arrows. The type of habitat where the species is most likely to be found in Ontario is described. Species distributions are denoted by two-letter codes that correspond to the five broad regions of Ontario shown on the title page of this guide (NW = northwestern, NE = northeastern, CE = central, SW = southwestern, SE = southeastern). Distributions were determined using Clarke (1981), the database of the NHIC, and the National Water Research Institute's Lower Great Lakes Unionid Database, which currently contains over 8000 records dating from 1860 to 2004. Finally, the SRANK, COSEWIC status designation and OMNR status designation are given for each species. These ranks were correct at the time of publication of the guide but are subject to change. Important terms used throughout the guide are defined in the Glossary (see page 59).

The rear cover of the guide presents a visual index to the species, which is intended to narrow your search and help you quickly identify specimens in the field. The page number next to each image corresponds to the page on which the species is profiled. The coloured tabs included on each species' page indicate the subfamily as shown on both the rear cover and the front index. A ruler is also included for field measurements.

Freshwater Mussel Shell Structure

External View of Right Valve

Internal View of Left Valve

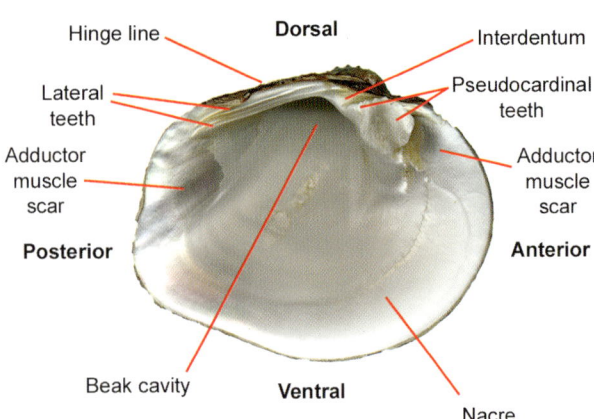

Reference Information

Guides to the freshwater mussels of Connecticut, Kentucky, Maine, Minnesota, New York, Tennessee, the Upper Mississippi River and the Midwest were consulted during the preparation of this guide. The guide was inspired by Arthur H. Clarke's book on the freshwater molluscs of Canada that was published almost 25 years ago. Some of the most useful guides for Ontario species, as well as several other important references, are listed below. "The Freshwater Mussels of Maine" provides in-depth coverage of a wide range of topics related to mussel classification, biology and conservation in a very readable format and is highly recommended for anyone interested in learning more about these fascinating creatures. The most complete source of information on freshwater mussels in North America is the web site of the Freshwater Mollusk Conservation Society (http://ellipse.inhs.uiuc.edu/FMCS).

Clarke, A.H. 1981. The Freshwater Molluscs of Canada. National Museums of Canada, Ottawa. 446 pp. ISBN 0-660-00022-9.

Cummings, K.S. and C.A. Mayer. 1992. Field Guide to Freshwater Mussels of the Midwest. Illinois Natural History Survey Manual 5. 194 pp. ISBN 1-882932-00-5. (Also available on-line at www.inhs.uiuc.edu/cbd/collections/mollusk/fieldguide.html).

Nedeau, E.J., M.A. McCollough and B.I. Swartz. 2000. The Freshwater Mussels of Maine. 118 pp. Available from the Maine Department of Inland Fisheries and Wildlife, Information Center, 41 State House Station, Augusta, Maine, 04333-0041, USA.

Parmalee, P.W. and A.E. Bogan. 1998. The Freshwater Mussels of Tennessee. University of Tennessee Press, Knoxville, Tennessee. 384 pp. ISBN 1-57233-013-9.

Strayer, D.L. and D.R. Smith. 2003. A Guide to Sampling Freshwater Mussel Populations. American Fisheries Society, Monograph 8, Bethesda, Maryland. 103 pp. ISBN 1-888569-50-6.

Strayer, D.L. and Jirka Strayer. 1997. The Pearly Mussels of New York State. New York State Museum Memoir 26. The New York State Museum, Albany, New York. 113 pp + 27 plates. ISBN 1-55557-155-7

Turgeon, D.D., J.F. Quinn, Jr., A.E. Bogan, E.V. Coan, F.G. Hochberg, W.G. Lyons, P.M. Mikkelsen, R.J. Neves, C.F.E. Roper, G. Rosenberg, B. Roth, A. Scheltema, F.G. Thompson, M. Vecchione and J.D. Williams. 1998. Common and Scientific Names of Aquatic Invertebrates from the United States and Canada: Mollusks, 2nd Edition. American Fisheries Society Special Publication 26. 526 pp. ISBN 1-888569-01-8.

Threeridge *Amblema plicata*

115 mm (max. 200)

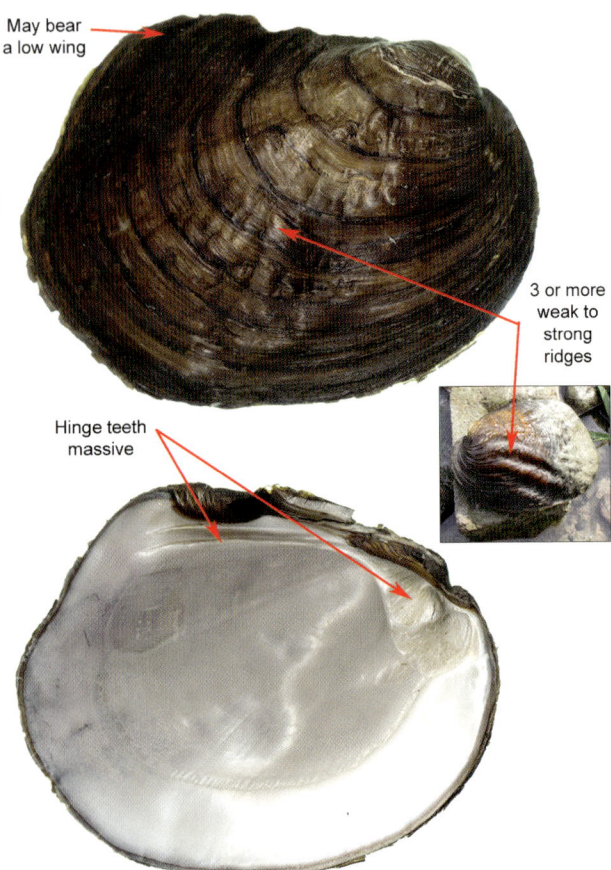

May bear a low wing

3 or more weak to strong ridges

Hinge teeth massive

Description: Shell thick, oval to quadrate, compressed, usually with 3 or more weak to strongly developed rounded ridges running across face of shell; dorsal margin may extend into a low wing; shell surface (periostracum) brown to blackish and without rays, may be light brown and without ridges in juveniles. Beaks slightly elevated above hinge line, close to anterior end; beak sculpture several concentric single-looped ridges visible only in juveniles. Nacre white, iridescent posteriorly. Hinge teeth massive: pseudocardinals triangular, deeply serrated; laterals moderately long, straight to slightly curved.

Habitat: Small streams to large rivers and reservoirs in mud, sand or gravel in areas with little to swift current.

Distribution: SW **SRANK**: S4 **COSEWIC**: None **OMNR**: None

Purple Wartyback *Cyclonaias tuberculata*   **85 mm** (max. 175)

Posterior two-thirds of shell covered in prominent nodules

Live specimens showing variation in shell shape

Interdentum wide and flat

Nacre purple to deep purple

Description: Shell thick, quadrate to circular, compressed; anterior and ventral margins rounded, posterior end squared off by a wing behind the beak; shell surface yellowish- or reddish-brown to dark brown, posterior two-thirds of shell covered with prominent nodules that become ridges on the dorsal wing. Beaks low; beak sculpture numerous fine zigzag ridges. Nacre purple to deep purple. Pseudocardinal teeth massive and serrated; lateral teeth thick and short; area between pseudocardinals and laterals (interdentum) wide and flat.
Habitat: Small to large rivers in gravel or mixed sand and gravel in areas with moderate to swift current.
Distribution: SW   **SRANK**: S3   **COSEWIC**: Candidate   **OMNR**: None

Eastern Elliptio *Elliptio complanata* **80 mm** (max. 125)

Pronounced posterior ridge

Shell surface rough or cloth-like

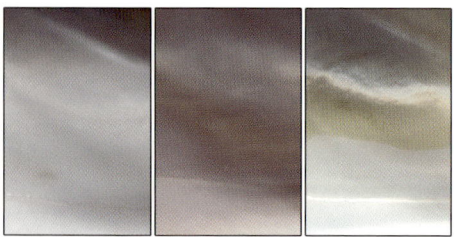

Nacre colour usually purple, may be rose or salmon

Description: Shell shape highly variable but usually moderately thick, trapezoidal, compressed, with a pronounced posterior ridge; shell surface rough or cloth-like, tan or brownish and sometimes with fine green rays in juveniles, dark brown to black in adults. Beaks nearly even with hinge line; beak sculpture concentric U-shaped ridges. Nacre usually purple, may be rose or salmon coloured. Pseudocardinal teeth triangular with rough surfaces; lateral teeth long, narrow and nearly straight.

Habitat: Virtually any permanent pond, lake, stream or river in all types of substrate including gravel, sand, clay and mud.

Distribution: SE, CE, NE **SRANK**: S5 **COSEWIC**: None **OMNR**: None

Elephantear *Elliptio crassidens*   Up to 150 mm

Posterior ridge prominent and sharply angled

Hinge teeth heavy, roughened

Shell shape variable

Nacre usually light purple

Description: Shell very thick, triangular, elongate; anterior end rounded, posterior end pointed, ventral margin straight or curved; posterior ridge prominent and sharply angled; shell surface rough, reddish-brown to black, rayless. Beaks large, flattened, slightly elevated above hinge line; beak sculpture 2-3 coarse loops parallel to growth lines, seldom visible. Nacre usually light purple, sometimes pink or white. Hinge teeth heavy, roughened: pseudocardinals triangular, divergent in left valve; laterals short, wide and straight. This species has not been confirmed to occur in Ontario.
Habitat: Large rivers in sand and coarse gravel often with a high percentage of mud in areas with strong current.
Distribution: SE, CE **SRANK**: SU **COSEWIC**: Candidate **OMNR**: None

Spike *Elliptio dilatata* **95 mm** (max. 135)

Shell elongated

Beak sculpture 4-5 heavy curved bars

Nacre usually dark purple, may be white or salmon

Description: Shell moderately thick, elongate, somewhat inflated; anterior end rounded, posterior end pointed, ventral margin straight or curved but arched in old specimens; shell surface of juveniles greenish-brown with faint rays, adults dark brown to black. Beaks flattened, barely extending above hinge line; beak sculpture 4-5 rather heavy curved bars. Nacre usually dark purple, may be white or salmon. Pseudocardinal teeth strong, serrated; lateral teeth moderately long, thick.

Habitat: Small streams to large rivers and occasionally lakes, preferring substrates of coarse sand and gravel with moderately strong current.

Distribution: SW, SE **SRANK**: S5 **COSEWIC**: None **OMNR**: None

Wabash Pigtoe *Fusconaia flava*   **60 mm** (max. 140)

Shallow furrow in front of a well-defined posterior ridge

Shell light to dark brown with a satinlike lustre

Beaks extend above hinge line, more central than in Round Pigtoe

Beak sculpture 3-5 weak bars

Description: Shell thick, quadrate to triangular, with a wide shallow furrow in front of a well-marked posterior ridge; anterior end rounded, posterior end bluntly pointed, ventral margin slightly arched posteriorly; shell surface light to dark brown with a satin-like lustre, faint green rays occasionally visible in juveniles. Beaks extend above hinge line, more central than in the Round Pigtoe; beak sculpture 3-5 weak bars most visible on the posterior ridge. Nacre white or tinged with salmon. Pseudocardinal teeth large, serrated; lateral teeth heavy, short.
Habitat: Typically medium-sized to large rivers but also small creeks and the Great Lakes in mud, sand or gravel.
Distribution: SW **SRANK**: S2S3 **COSEWIC**: Candidate **OMNR**: None

Round Pigtoe *Pleurobema sintoxia* **70 mm** (max. 130)

Shell typically a reddish-brown

Beaks full, directed forward and close to anterior end of shell

Nacre white, pink or rose

Live specimens showing typical colouration

Description: Shell thick, triangular to oval, compressed to inflated, highly variable; anterior end rounded, posterior end bluntly pointed, dorsal margin straight to slightly curved, ventral margin curved to concave; shell surface typically a deep reddish-brown, lighter in juveniles and blackish in old adults. Beaks full, directed forward and close to anterior end of shell; beak sculpture 2-3 small ridges visible only in juveniles. Nacre white, pink or rose. Pseudocardinal teeth heavy, serrated; lateral teeth short and straight.
Habitat: Medium-sized to large rivers with moderate flows in mixed substrates of boulder, cobble, gravel, sand and mud.
Distribution: SW **SRANK**: S2S3 **COSEWIC**: Endangered
OMNR: Under consideration

Pimpleback *Quadrula pustulosa* **70 mm** (max. 100)

Posterior two-thirds of shell covered with irregular nodules

Broad green ray near the beak

Juvenile

Beak sculpture indistinct

Shell shape circular to quadrate

Description: Shell thick, circular to somewhat quadrate, moderately inflated; anterior end rounded, posterior end truncated, ventral margin curved; shell surface yellowish-brown to chestnut-brown in adults, lighter in juveniles and with a broad green ray near the beak, posterior two-thirds of shell typically covered with irregular nodules (rarely absent). Beaks full, directed forward; beak sculpture 3-4 indistinct ridges. Nacre white, iridescent posteriorly. Pseudocardinal teeth large, serrated; lateral teeth short, thick and straight to slightly curved.
Habitat: Medium-sized to large rivers and reservoirs in substrates of coarse gravel, sand and silt in flowing water.
Distribution: SW **SRANK**: S3 **COSEWIC**: Candidate **OMNR**: None

Mapleleaf *Quadrula quadrula* **90 mm** (max. 130)

Two rows of nodules separated by a shallow furrow

Posterior end squared with a small wing

Juvenile

Beak sculpture 2 rows of nodules that continue down the shell

Description: Shell thick, quadrate, with 2 rows of few to many nodules separated by a shallow furrow; anterior end rounded, posterior end squared with a small wing, ventral margin concave; shell surface yellowish-green with faint rays in juveniles, yellowish-brown to dark brown in adults. Beaks small, extending above hinge line; beak sculpture 2 rows of small, crowded nodules that continue down the shell surface. Nacre pearly white, iridescent posteriorly. Pseudocardinal teeth heavy, serrated, divergent in left valve; lateral teeth high, moderately long, finely striated.

Habitat: Medium-sized to large rivers and reservoirs where currents are slow to moderate in soft or coarse substrates.

Distribution: SW **SRANK**: S3 **COSEWIC**: Under review **OMNR**: None

Elktoe *Alasmidonta marginata* **70 mm** (max. 140)

Shell green to yellowish-green with dark green rays and speckles

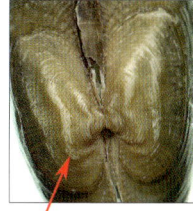

Beak sculpture 3-4 heavy double-looped bars

Lateral teeth absent

Beautifully marked live specimen

Description: Shell thin to moderately thick, triangular; anterior end rounded, posterior end bluntly pointed, ventral margin straight or slightly concave; posterior ridge sharp, posterior slope flat with numerous fine ridges; shell surface green to yellowish-green with numerous dark green rays and speckles. Beaks large, central, extend above hinge line; beak sculpture 3-4 heavy double-looped bars. Nacre bluish-white or white. Pseudocardinal teeth small, nodular, smooth; lateral teeth absent, appearing as a thickened hinge line. Live animals have a bright orange foot.
Habitat: Small streams to medium-sized rivers in gravel or mixed sand and gravel in riffles; usually deeply buried in the substrate.
Distribution: SW, SE, CE **SRANK**: S3 **COSEWIC**: Candidate
OMNR: None

Triangle Floater *Alasmidonta undulata* **50 mm** (max. 75)

Beak sculpture very heavy with ridges extending far out onto shell

Shell inflated

Pseudocardinal teeth strong and deeply grooved

Description: Shell thick at anterior end and thin posteriorly, triangular, inflated; ventral margin rounded; shell surface yellowish-green to greenish-brown with numerous green rays in juveniles, black and rayless in old adults. Beaks full, elevated above hinge line; beak sculpture very heavy, composed of several single-looped ridges extending far out onto the shell. Nacre white or pinkish anteriorly, bluish-pink posteriorly. Pseudocardinal teeth strong and deeply grooved; lateral teeth vestigial or absent.

Habitat: Mainly streams and rivers but also lakes and ponds in sand and gravel substrates; can tolerate standing water.

Distribution: SE, CE **SRANK**: S4 **COSEWIC**: Candidate **OMNR**: None

Slippershell Mussel *Alasmidonta viridis* **30 mm** (max. 50)

Shell greyish-green with numerous wavy green rays

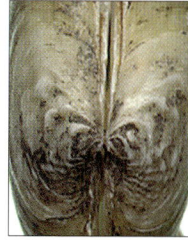

Beak sculpture 6-8 irregular loops

Lateral teeth poorly developed

Live specimens showing small size

Posterior end squared

Description: Shell relatively thin, trapezoidal; anterior end rounded, posterior end squared, ventral margin straight or slightly convex; posterior ridge rounded; shell surface a characteristic greyish-green, usually with numerous wavy green rays. Beaks extend slightly above hinge line; beak sculpture 6-8 irregular heavy loops. Nacre white or bluish-white, iridescent posteriorly. Pseudocardinal teeth moderate in size, triangular, serrated; lateral teeth poorly developed or reduced to a swelling along the hinge line.
Habitat: Small streams and the headwaters of large rivers in sand, fine gravel or mud.
Distribution: SW **SRANK**: S3 **COSEWIC**: Candidate **OMNR**: None

Cylindrical Papershell *Anodontoides ferussacianus* **55 mm**
(max. 95)

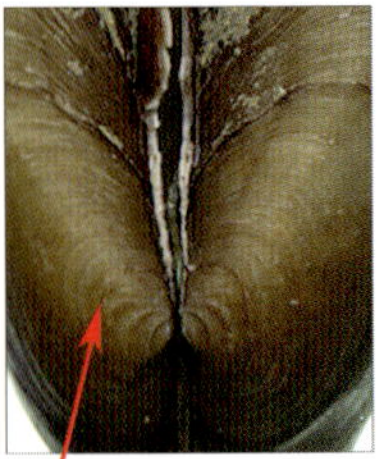

Growth rings prominent

Beak sculpture 3-4 very fine V-shaped ridges not parallel to growth rings

Shell sometimes has fine green rays

Hinge teeth absent

Description: Shell thin, elliptical, elongate; anterior end rounded, posterior end bluntly pointed, dorsal margin nearly straight, ventral margin often pinched in the middle; shell surface smooth, shiny, greenish or brownish and sometimes with fine green rays, growth rings prominent. Beaks small, extend slightly above hinge line; beak sculpture 3-4 very fine V-shaped ridges not parallel to growth rings. Nacre bluish-white, silvery. Hinge teeth absent except for a slight thickening in front of the beak.
Habitat: Small, slow-moving streams and the headwaters of large streams in silt or mud or sometimes sand; occasionally in lakes or large rivers.
Distribution: SW, SE, CE, NE, NW **SRANK**: S4 **COSEWIC**: None
OMNR: None

White Heelsplitter *Lasmigona complanata* **140 mm** (max. 200)

Juvenile showing the prominent dorsal wing

Beak sculpture up to 8 heavy double-looped bars

Lateral teeth reduced to a thickening or absent

Description: Shell very thin, circular, with prominent dorsal wing in juveniles, oval and thicker in adults, compressed; anterior end rounded, posterior end rounded or bluntly pointed, ventral margin slightly curved; shell surface brown with a few rays in juveniles, blackish-brown and rayless in adults. Beaks small, even with hinge line; beak sculpture up to 8 heavy, double-looped bars resembling a butterfly. Nacre white, iridescent. Pseudocardinal teeth large, irregularly shaped, low, flattened; lateral teeth reduced to a thickening along the hinge line or absent.
Habitat: Small streams to medium-sized rivers, lakes and reservoirs in quiet water on sandy or muddy substrates; tolerant of disturbance.
Distribution: SW, NW **SRANK**: S4 **COSEWIC**: None **OMNR**: None

Creek Heelsplitter *Lasmigona compressa* 80 mm (max. 115)

Small dorsal wing

Beak sculpture 5-8 double-looped ridges

Lateral teeth thin and undeveloped near beaks

Prominent interdental tooth in the left valve

Description: Shell relatively thin, trapezoidal, compressed, with a small dorsal wing; anterior end broadly rounded, posterior end bluntly pointed and squared at tip, ventral margin curved; shell surface smooth, greenish- or yellowish-brown to brown, extensively but not prominently rayed. Beaks small, extending slightly above hinge line; beak sculpture 5-8 double-looped ridges, obvious. Nacre white, sometimes cream or salmon near beak cavities. Pseudocardinal teeth smooth, low; lateral teeth thin, undeveloped near beaks; very prominent interdental tooth in the left valve.
Habitat: Small streams and the headwaters of small to medium-sized rivers in fine gravel or sand.
Distribution: SW, SE, CE, NE, NW **SRANK**: S5 **COSEWIC**: None
OMNR: None

Flutedshell *Lasmigona costata* **105 mm** (max. 175)

Posterior slope with 10-20 heavy ridges

Beak sculpture 3-4 heavy double-looped ridges

Prominent interdental tooth

Description: Shell moderately thick, oval, elongate, compressed; anterior end rounded, posterior end rounded or bluntly pointed, dorsal and ventral margins fairly straight; posterior slope with 10-20 heavy ridges crossing the growth lines; shell surface greenish-brown to dark brown with numerous narrow green rays becoming obscure with age. Beaks low; beak sculpture 3-4 heavy double-looped ridges parallel to hinge line. Nacre white with pink or salmon tints. Pseudocardinal teeth moderately strong; lateral teeth vestigial or absent; prominent interdental tooth.
Habitat: Small to large rivers in a variety of substrates but preferring coarse gravel and sand in areas with moderately strong current.
Distribution: SW, SE, CE, NW **SRANK**: S5 **COSEWIC**: None
OMNR: None

Eastern Floater *Pyganodon cataracta*   100 mm (max. 150)

Shell surface smooth and shiny, yellowish-green, green or greenish-brown, often with faint green rays

Beak sculpture 6-8 double-looped bars of uniform height

Hinge teeth absent

Description: Shell uniformly thin and fragile, elliptical, inflated; anterior end rounded, posterior end drawn out and pointed, ventral margin rounded, dorsal margin straight or curved slightly upward; shell surface smooth and shiny, yellowish-green, green or greenish-brown, often with faint green rays. Beaks slightly inflated and extend above hinge; beak sculpture 6-8 double-looped bars of uniform height. Nacre bluish-white or silvery, sometimes tinged with yellow. Hinge teeth absent.

Habitat: A wide range of habitats including small streams, rivers, ponds and lakes in quiet protected waters in mud or silt, less frequently in sand or gravel.

Distribution: SE **SRANK**: S2 **COSEWIC**: None **OMNR**: None

Giant Floater *Pyganodon grandis*   95 mm (max. 160)

Shell shape highly variable

Beak sculpture 4-5 nodulous, double-looped ridges

Hinge teeth absent, but hinge line may be thickened

Description: Shell shape highly variable but usually oval to elongate, thin and compressed in juveniles, somewhat thicker and inflated in adults; anterior end broadly rounded, posterior end bluntly pointed, ventral margin straight or curved; shell surface smooth, yellowish-green or greenish-brown with faint rays in juveniles, dark green or dark brown in older adults. Beaks inflated, extending well above hinge line; beak sculpture 4-5 nodulous, double-looped ridges. Nacre silvery-white often tinged with cream, salmon or pink. Hinge teeth absent but hinge line may be thickened.
Habitat: Small streams to large rivers in backwaters with little or no current in clay, silt or mud and in lakes, wetlands, ponds and reservoirs.
Distribution: SW, SE, CE, NE, NW **SRANK**: S5 **COSEWIC**: None
OMNR: None

Salamander Mussel *Simpsonaias ambigua* **25 mm** (max. 40)

Shell elliptical and elongate

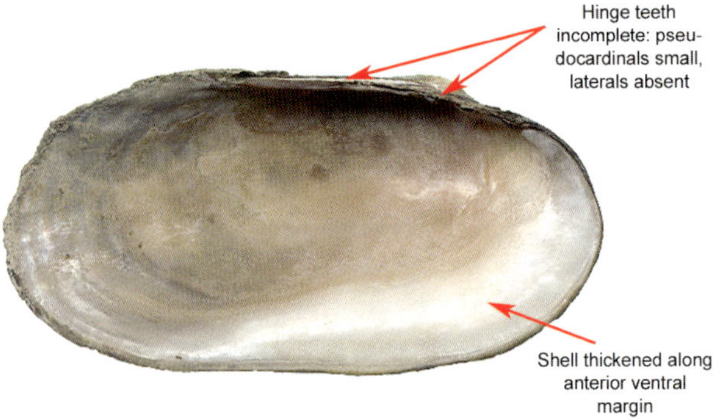

Hinge teeth incomplete: pseudocardinals small, laterals absent

Shell thickened along anterior ventral margin

Live animals found under this flat rock

Description: Shell elliptical, elongate, thin and fragile except thickened along anterior ventral margin, slightly inflated posteriorly (females); anterior and posterior ends rounded, dorsal and ventral margins parallel, straight; shell surface dull, yellowish-tan to dark brown, rayless. Beaks small, slightly elevated above hinge line, close to anterior end; beak sculpture 4-5 fine ridges drawn up into an inverted V, obscure. Nacre white, often tinged with yellow or salmon. Hinge teeth incomplete: pseudocardinal teeth small, flattened; lateral teeth absent.
Habitat: Medium-sized to large rivers usually buried in sand or silt under flat rocks or rock ledges used as shelters by its host, the Mudpuppy; sometimes found in mud or gravel bars; also known from Lake Erie.
Distribution: SW **SRANK**: S1 **COSEWIC**: Endangered **OMNR**: END

Creeper *Strophitus undulatus*   **70 mm** (max. 100)

Shell surface smooth, greenish-brown with fine green rays in juveniles

Juvenile

Beak sculpture 3-4 heavy, concentric bars

Pseudocardinal teeth vestigial, indicated by a thickening in front of the beak

Description: Shell moderately thin, oval to trapezoidal, somewhat compressed; anterior end rounded, posterior end bluntly pointed, ventral margin curved to slightly concave; shell surface smooth, greenish-brown with fine green rays in juveniles, dark brown to black in adults. Beaks central, slightly elevated above hinge line; beak sculpture 3-4 heavy, concentric bars. Nacre bluish-white with cream or salmon near beak cavities. Pseudocardinal teeth vestigial, indicated by a thickening of the hinge line in front of the beak; lateral teeth absent. Live animals often have an orange foot.

Habitat: Small to medium-sized streams or occasionally large rivers in mud, sand or fine gravel in a range of flow conditions.

Distribution: SW, SE, CE, NW **SRANK**: S5 **COSEWIC**: None **OMNR**: None

Paper Pondshell *Utterbackia imbecillis* 70 mm (max. 90)

Shell elongate and extremely thin

Beak sculpture 5-6 fine ridges

Beaks even with hinge line

Hinge teeth absent

Description: Shell elongate and extremely thin, juveniles greatly compressed, adults inflated; anterior end rounded, posterior end pointed, dorsal and ventral margins relatively straight, sometimes with a small wing behind the beak; shell surface smooth, shiny, green to greenish-brown, covered with numerous fine rays that become broad on the posterior ridge and slope. Beaks even with hinge line; beak sculpture 5-6 fine ridges forming indistinct double loops. Nacre bluish-white, iridescent posteriorly. Hinge teeth absent.

Habitat: Ponds, lakes and backwaters of streams and rivers in muddy to somewhat sandy substrates; rarely found in gravel.

Distribution: SW, SE **SRANK**: S2 **COSEWIC**: Candidate **OMNR**: None

Mucket *Actinonaias ligamentina*

130 mm (max. 190)

Beaks extend slightly above hinge line

Beak sculpture usually obscure

Rayless form

Shell very thick

Description: Shell very thick, elliptical, moderately compressed to slightly inflated; anterior end rounded, posterior end bluntly pointed, ventral margin broadly curved; shell surface smooth except for raised growth rings, yellowish-brown to dark brown, often with broad green rays. Beaks extend slightly above hinge line; beak sculpture a few faint double-looped ridges, usually obscure. Nacre white, rarely pinkish, iridescent posteriorly. Pseudocardinal teeth moderately heavy, serrated; lateral teeth thick and curved.

Habitat: Medium-sized to large rivers in substrates ranging from cobble and gravel in riffles with strong currents to quiet water with coarse gravel, sand or mud.

Distribution: SW, SE **SRANK**: S3 **COSEWIC**: Candidate **OMNR**: None

Northern Riffleshell *Epioblasma torulosa rangiana*   **50 mm**
(max. 75)

♂

Shell brightly coloured with crowded dark green rays

♀

Posterior end in males indented and with a deep furrow

Ventral margin rounded and distended in females

♂

Description: Sexually dimorphic. Shell thick anteriorly, much thinner posteriorly, quadrate in males, oval in females, inflated; anterior end rounded, posterior end indented and with a deep furrow in males, broadly rounded and distended along ventral margin in females; shell surface bright yellowish-green to yellowish-orange with many crowded broad or narrow dark green rays. Beaks extend above hinge line; beak sculpture a few weak single-looped bars, obscure. Nacre white, iridescent posteriorly. Pseudocardinal teeth triangular, roughened, divergent in left valve; lateral teeth moderately long, straight.
Habitat: Small to large rivers in coarse sand and gravel with some cobble to firmly packed fine gravel in shallow riffle areas with swift current.
Distribution: SW **SRANK**: S1 **COSEWIC**: Endangered **OMNR**: END

Snuffbox *Epioblasma triquetra* 50 mm (max. 75)

Description: Sexually dimorphic. Shell thick, triangular, posterior ridge prominent; anterior end rounded, posterior end truncated, ventral margin curved in males, centrally pinched in females; posterior slope flat and ribbed; female shell smaller than male, highly inflated along posterior ridge; shell surface smooth, yellow to yellowish-green, covered with dark green rays and blotches resembling dripping paint. Beaks large, extending above hinge line; beak sculpture 3-4 faint, double-looped bars. Nacre white. Pseudocardinal teeth large, serrated; lateral teeth short, thick.
Habitat: Medium-sized to large rivers in shallow riffles with clear, swift-flowing water and firm coarse sand and gravel substrates; usually buried.
Distribution: SW **SRANK**: S1 **COSEWIC**: Endangered **OMNR**: END

Plain Pocketbook *Lampsilis cardium*   **95 mm** (max. 155)

Shell surface smooth and yellow to yellowish-green with widely separated green rays

Shell very inflated

Beak sculpture 5-6 double-looped bars

Beaks swollen and extend well above hinge line

Description: Sexually dimorphic. Shell moderately thick to thick, oval, very inflated; anterior end sharply rounded, posterior end bluntly pointed in males, squared off in females; shell surface smooth and yellow to yellowish-green with widely separated dark green rays, but old specimens may be rayless. Beaks swollen and extend well above hinge line; beak sculpture 5-6 double-looped bars, the last 2 or 3 prominent. Nacre white, sometimes tinged with pink. Pseudocardinal teeth prominent, compressed, directed forward; lateral teeth moderately long and slightly curved.

Habitat: Small streams to large rivers in gravel, sand or mud; prefers moderate to strong current.

Distribution: SW, SE, CE   **SRANK**: S4   **COSEWIC**: None   **OMNR**: None

Wavyrayed Lampmussel *Lampsilis fasciola* **60 mm**
(max. 100)

Shell margin evenly rounded

Shell surface densely covered with wavy green interrupted rays

Lateral teeth short, thick, nearly straight

Description: Sexually dimorphic. Shell moderately thick, oval, inflated; shell margin evenly rounded, expanded posteriorly in females; shell surface smooth, yellow, densely covered with wavy green rays of varying widths that are often interrupted at growth lines. Beaks full, extending slightly above hinge line; beak sculpture several indistinct, fine wavy ridges. Nacre white or bluish-white, iridescent posteriorly. Pseudocardinal teeth large, coarsely serrated; lateral teeth short, thick, nearly straight.

Habitat: Small to medium-sized rivers with steady flows and clear water in and around riffle areas in gravel or sand often stabilized with cobble or boulders.

Distribution: SW **SRANK**: S1 **COSEWIC**: Endangered **OMNR**: END

Eastern Lampmussel *Lampsilis radiata* 80 mm (max. 105)

Shell rough with numerous, prominent, mostly broad green rays

Beaks barely extend above hinge line

Beak sculpture 6 double-looped bars visible only in juveniles

Description: Sexually dimorphic. Shell moderately thick, oval to elliptical, somewhat compressed; anterior end rounded, ventral margin curved, posterior end pointed in males, rounded and expanded in females; shell surface rough, yellowish-green, greenish-brown or blackish, covered with numerous, prominent, mostly broad green rays. Beaks barely extend above hinge line; beak sculpture 6 double-looped bars visible only in juveniles. Nacre white, bluish-white or pink. Pseudocardinal teeth medium-sized, triangular; lateral teeth long, relatively straight.

Habitat: Streams, rivers, ponds and lakes usually in gravel or sand but occasionally mud; often very abundant.

Distribution: SE, CE **SRANK**: S4 **COSEWIC**: None **OMNR**: None

Fatmucket *Lampsilis siliquoidea*   **70 mm** (max. 155)

Shell surface smooth, yellow or brownish with widely spaced narrow green rays

Female shell smaller and very inflated

Beak sculpture 6-10 double-looped bars

Description: Sexually dimorphic. Shell thick, elliptical, moderately compressed (males), very inflated and smaller (females); anterior end rounded, ventral margin straight or slightly rounded, posterior end bluntly pointed (males), squared (females); shell surface smooth, yellow or brownish with widely spaced narrow green rays. Beaks extend slightly above hinge line; beak sculpture 6-10 double-looped bars. Nacre white with bluish or pinkish tinge, iridescent posteriorly. Pseudocardinal teeth medium-sized, compressed; lateral teeth narrow, straight to slightly curved.

Habitat: Small to medium-sized streams in clay, mud, sand or gravel often in very shallow waters along the margins of pools; also occurs in lakes; frequently abundant.

Distribution: SW, SE, CE, NE, NW **SRANK**: S5 **COSEWIC**: None
OMNR: None

Fragile Papershell *Leptodea fragilis*   **100 mm** (max. 185)

Shell smooth except for rough growth rings, pale yellow or tan with faint green rays

Low dorsal wing usually present in juveniles and young adults

Young adult

Pseudocardinal teeth small, rudimentary

Description: Sexual dimorphism is subtle. Shell thin, brittle, oval to elliptical, moderately compressed; anterior and posterior ends rounded (distended in females), ventral margin straight or slightly curved, low dorsal wing usually present in younger specimens; shell surface smooth except for rough growth rings, pale yellow or tan with faint green rays. Beaks flattened, slightly elevated above hinge line; beak sculpture 3-4 faint double-looped bars. Nacre white or pinkish-white, highly iridescent. Pseudocardinal teeth small, rudimentary; lateral teeth long, thin.

Habitat: Streams and rivers of all sizes and reservoirs in a variety of substrates and flow conditions but prefers firm sand and mud in slow current; always firmly anchored.

Distribution: SW, SE **SRANK**: S4 **COSEWIC**: Candidate **OMNR**: None

Eastern Pondmussel *Ligumia nasuta*   70 mm (max. 100)

Posterior end tapering to a blunt point

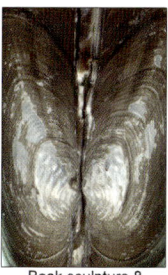

Beak sculpture 8 fine double-looped bars

Shell thin but strong, narrow and elongate

Nacre purple, pink or silvery-white

Description: Sexual dimorphism is subtle. Shell thin but strong, narrow and elongate; anterior end rounded, posterior end tapering to a blunt point, females distended along posterior ventral margin; posterior ridge well marked; shell surface olive-green, brown or blackish, often with narrow rays. Beaks project slightly above hinge line; beak sculpture about 8 fine, double-looped bars. Nacre purple, pink or silvery-white. Pseudocardinal and lateral teeth sharp, delicate.
Habitat: Ponds, lakes and slow-moving rivers, especially the lower reaches of large rivers and some nearshore areas in the lower Great Lakes in mud or sand in quiet waters.
Distribution: SW, SE **SRANK**: S2S3 **COSEWIC**: Candidate **OMNR**: None

Black Sandshell *Ligumia recta* **150 mm** (max. 205)

Shell thick and elongate

Large female

Shell surface black with obscure rays in adults

Nacre usually white tinged with pink dorsally

Description: Sexually dimorphic. Shell thick, elongate, moderately inflated; anterior end rounded, posterior end pointed in males and distended in females, dorsal and ventral margins straight and parallel; shell surface shiny and dark green with numerous rays in juveniles, black with obscure rays in adults. Beaks low, nearly even with hinge line; beak sculpture 3-5 indistinct double-looped bars. Nacre usually white tinged with pink dorsally, may be pink or purple. Pseudocardinal teeth medium-sized, triangular, elevated; lateral teeth long, straight, elevated.

Habitat: Medium-sized to large rivers in riffles or raceways in gravel or firm sand and occasionally mud; often in very shallow water.

Distribution: SW, SE, NW **SRANK**: S3 **COSEWIC**: Candidate
OMNR: None

Threehorn Wartyback *Obliquaria reflexa*   **40 mm** (max. 55)

Posterior slope ribbed

2-5 large knobs extending from beak to ventral margin and alternating in position between valves

Pseudocardinal teeth strong and deeply serrated

Description: Shell thick, circular to triangular, inflated, with 2-5 large knobs extending from beak to ventral margin and alternating in position between valves; anterior end rounded, posterior end bluntly pointed, ventral margin curved anteriorly and slanted upward posteriorly; posterior slope ribbed; shell surface green, tan or brown and sometimes with thin rays. Beaks elevated, curved inward; beak sculpture a miniature version of adult shell. Nacre white, iridescent posteriorly. Pseudocardinal teeth strong and deeply serrated; lateral teeth thick and short.

Habitat: Typical of large rivers with stable gravel, sand and mud substrates and moderate current, but also occurs in shallow embayments and reservoirs with almost no current.

Distribution: SW **SRANK**: S1 **COSEWIC**: Candidate **OMNR**: None

Hickorynut *Obovaria olivaria*

55 mm (max. 80)

Beaks inflated, directed forward and very close to anterior end of shell

Shell thick, almost perfectly oval

Specimen with rays

Pseudocardinal teeth heavy, roughened, parallel with lateral teeth

Description: Sexual dimorphism is subtle. Shell thick, almost perfectly oval, inflated; anterior and ventral margins rounded, posterior margin bluntly rounded (males) or broadly rounded (females); shell surface smooth, olive-green or yellowish-brown with faint rays in juveniles, dark brown in older specimens. Beaks inflated, directed forward and very close to anterior end of shell; beak sculpture 4-5 delicate double-looped bars, obscure. Nacre bright white, iridescent posteriorly. Pseudocardinal teeth heavy, roughened, parallel with lateral teeth; lateral teeth thick and long.
Habitat: Mainly large rivers, rarely in streams, on sand or mixed sand and gravel in deep water with good current; often found in gravel bars mid-river.
Distribution: SW, SE **SRANK**: S1 **COSEWIC**: Under review
OMNR: None

Round Hickorynut *Obovaria subrotunda*   **30 mm** (max. 65)

Shell distinctly lighter on posterior slope

Shell thick, almost perfectly circular

Beaks elevated well above hinge line, curved inward and centrally placed

Description: Sexual dimorphism is subtle. Shell thick, almost perfectly circular, inflated; all margins rounded except the posterior margin in females may be truncated; shell surface yellowish- or greenish-brown or brown, distinctly lighter on the posterior slope. Beaks elevated well above hinge line, curved inward and centrally placed; beak sculpture a few indistinct bars usually evident only in young specimens. Nacre silvery-white, iridescent posteriorly. Pseudocardinal teeth large, triangular, serrated; lateral teeth short, moderately thick and nearly straight.

Habitat: Deeper waters of medium-sized to large rivers with steady, moderate flows and sand and gravel substrates, but tolerates turbid water and some clay; also found in Lake St. Clair.

Distribution: SW **SRANK**: S1 **COSEWIC**: Endangered **OMNR**: END

Pink Heelsplitter *Potamilus alatus*   **110 mm** (max. 180)

Shell black in older specimens

Prominent dorsal wing in young specimens

Beak sculpture indistinct

Nacre purple or pinkish-purple, highly iridescent

Description: Shell oval, compressed, thin with a prominent dorsal wing in young specimens, moderately thick with a less pronounced wing in older adults; anterior end rounded, posterior end rounded or squared, ventral margin straight; shell surface dark green to dark brown with green rays in juveniles, becoming black with age. Beaks flattened, nearly even with hinge line; beak sculpture 3-4 narrow bars, indistinct. Nacre purple or pinkish-purple, highly iridescent. Pseudocardinal teeth relatively small, roughened; lateral teeth thin, elevated, slightly curved.

Habitat: Medium-sized and large rivers in mixed mud, sand and gravel; especially abundant in quiet backwaters.

Distribution: SW, SE, NW   **SRANK**: S3   **COSEWIC**: Candidate
OMNR: None

Kidneyshell *Ptychobranchus fasciolaris* **95 mm** (max. 125)

Shell surface usually covered with broad, interrupted green rays

Inside of female shell marked by a fold or groove

Juvenile

Lateral teeth thick, widely separated, spatulate in shape

Description: Shell thick, elliptical, moderately compressed, becoming humped (kidney-shaped) in older specimens; anterior end rounded, posterior end bluntly pointed, ventral margin straight to curved or arched; shell surface greenish-yellow or yellowish-brown and usually covered with broad, interrupted green rays. Beaks approximately even with hinge line; beak sculpture several indistinct, wavy ridges. Nacre white; inside of female shell marked by a fold or groove. Pseudocardinal teeth thick, serrated; lateral teeth thick, widely separated, spatulate in shape.

Habitat: Large streams and small rivers in firmly packed coarse gravel and sand in moderately strong current; often found near beds of aquatic vegetation.

Distribution: SW **SRANK**: S1 **COSEWIC**: Endangered **OMNR**: END

Lilliput *Toxolasma parvus*

25 mm (max. 35)

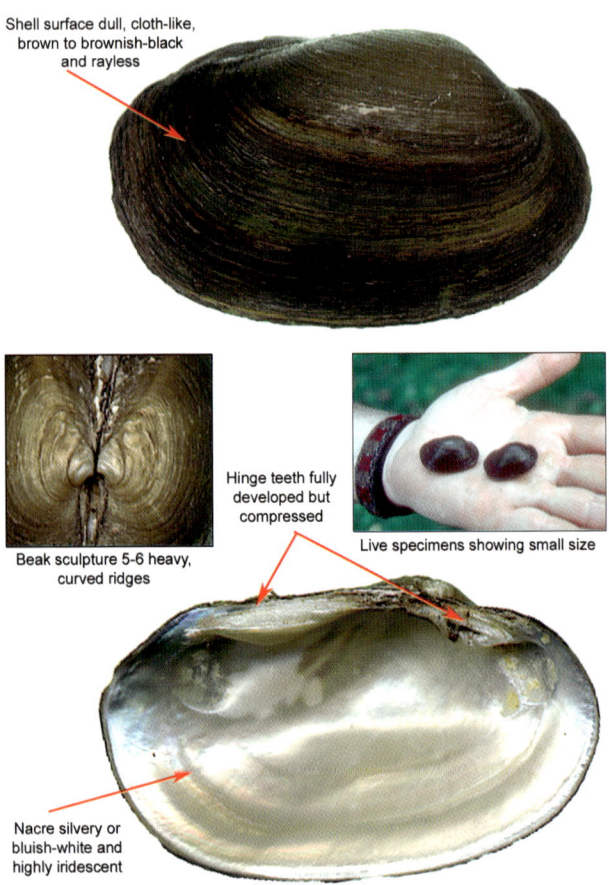

Shell surface dull, cloth-like, brown to brownish-black and rayless

Beak sculpture 5-6 heavy, curved ridges

Hinge teeth fully developed but compressed

Live specimens showing small size

Nacre silvery or bluish-white and highly iridescent

Description: Sexual dimorphism is subtle. Shell thick, elliptical, moderately inflated in males, oval and more inflated in females; anterior end rounded, posterior end rounded (males) or squared (females), ventral margin straight or slightly curved; shell surface dull, cloth-like, brown to brownish-black and rayless. Beaks inflated, slightly elevated above hinge line; beak sculpture 5-6 heavy, curved ridges. Nacre silvery or bluish-white and highly iridescent. Hinge teeth fully developed but compressed: pseudocardinal teeth thin, serrated; lateral teeth long, thin, straight.

Habitat: The lower reaches of large rivers and wetlands in backwater areas with little current and soft substrates of mud and sand.

Distribution: SW **SRANK**: S1 **COSEWIC**: Candidate **OMNR**: None

Fawnsfoot *Truncilla donaciformis*   35 mm (max. 45)

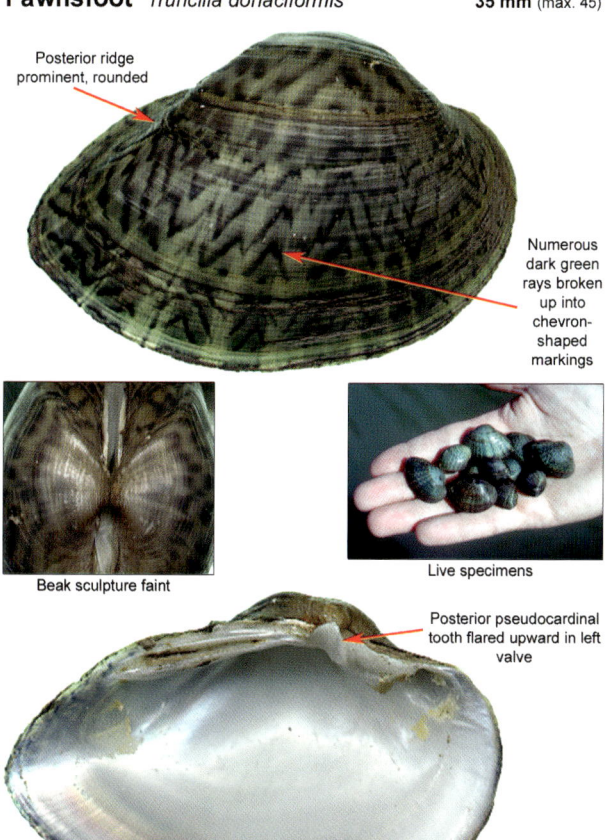

Posterior ridge prominent, rounded

Numerous dark green rays broken up into chevron-shaped markings

Beak sculpture faint

Live specimens

Posterior pseudocardinal tooth flared upward in left valve

Description: Shell moderately thick, oval to triangular; anterior end rounded, posterior end pointed, ventral margin curved; posterior ridge prominent, rounded; shell surface smooth, yellow to greenish-brown with numerous dark green rays broken up into large chevron-shaped markings. Beaks full, central, slightly elevated above hinge line; beak sculpture up to 8 faint, double-looped bars. Nacre bluish-white, iridescent posteriorly. Pseudocardinal teeth medium-sized, elevated, with posterior tooth in left valve flared upward; lateral teeth thin, long, straight to slightly curved.

Habitat: Large and medium-sized rivers in gravel, sand or mud in areas with little current.

Distribution: SW **SRANK**: S2 **COSEWIC**: Candidate **OMNR**: None

Deertoe *Truncilla truncata*

60 mm (max. 95)

- Posterior ridge sharply angled
- Shell usually covered in green rays composed of fine chevron markings
- Beak sculpture fine
- Beautifully marked live specimen
- Beaks full, curved inward and elevated well above hinge line

Description: Shell moderately thick, triangular, inflated; anterior end rounded, posterior end pointed, ventral margin curved or sometimes concave posteriorly; posterior ridge sharply angled; shell surface yellowish, greenish or brownish usually covered in green rays composed of fine chevron markings but occasionally rayless. Beaks full, curved inward, elevated well above hinge line; beak sculpture a few fine double-looped lines. Nacre white, sometimes pinkish, iridescent posteriorly. Pseudocardinal teeth strong, elevated, serrated; lateral teeth high, moderately long.

Habitat: Medium-sized to large rivers and shallow areas of the Great Lakes in mud, sand or gravel.

Distribution: SW **SRANK**: S3 **COSEWIC**: Candidate **OMNR**: None

Rayed Bean *Villosa fabalis*

20 mm (max. 40)

♂

Shell covered in dark green wide or narrow wavy rays

Live specimens

♀

Hinge teeth heavy

♂

Shell very thick

Description: Sexually dimorphic. Shell very thick, elliptical, moderately inflated in males and more so in females; anterior end rounded, posterior end pointed in males and rounded in females, dorsal and ventral margins curved; shell surface dark green to greenish-brown and covered in darker green wide or narrow wavy rays clearly evident except in old blackened specimens. Beaks extend slightly above hinge line; beak sculpture 5 faint, crowded, double-looped ridges. Nacre silvery-white, iridescent. Hinge teeth heavy: pseudocardinals triangular, low; laterals short and stout.

Habitat: Medium-sized streams to medium-sized rivers in sand and gravel; usually deeply buried among the roots of aquatic vegetation in and near riffles or along the river's edge.

Distribution: SW **SRANK**: S1 **COSEWIC**: Endangered **OMNR**: END

Rainbow *Villosa iris*

55 mm (max. 85)

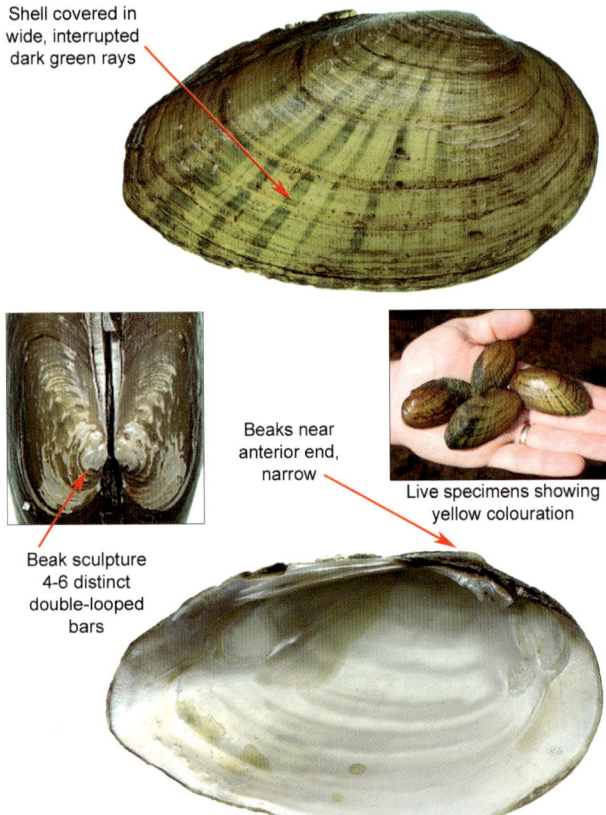

Shell covered in wide, interrupted dark green rays

Beaks near anterior end, narrow

Live specimens showing yellow colouration

Beak sculpture 4-6 distinct double-looped bars

Description: Sexual dimorphism is subtle. Shell relatively thin, elliptical, elongate, compressed (males) to moderately inflated (females); anterior end rounded, posterior end bluntly pointed (males) or rounded (females), dorsal and ventral margins straight or slightly curved; shell surface yellow or yellowish-green with wide, interrupted dark green rays that are more prominent posteriorly. Beaks near anterior end, narrow, extend slightly above hinge line; beak sculpture 4-6 distinct double-looped bars. Nacre white, iridescent posteriorly. Pseudocardinal teeth small, compressed; lateral teeth long, thin.

Habitat: Mainly small streams to small rivers in coarse sand or gravel substrates in or near riffles and along the edges of emergent vegetation in moderate to strong current.

Distribution: SW, SE **SRANK**: S2S3 **COSEWIC**: Under review
OMNR: None

Glossary

Adductor muscle: A large bundle of muscle fibres used to pull the two valves of the shell together.
Anterior end: The front end of the mussel where the foot is located.
Beak: The raised part of the dorsal margin of the shell representing the earliest period of shell growth; also known as the umbo.
Beak cavity: The hollow on the inside of the shell under the beak.
Beak sculpture: Raised loops, ridges or bars on the beak; often eroded in older specimens.
Calcareous: Made of calcium carbonate.
Compressed: Flattened or pressed together laterally.
Concentric: Having a common centre.
Conglutinate: A mass of glochidia bound together in a mucous matrix, often resembling prey items of the host fish.
Dorsal: The top part of the shell where the hinge is located.
Exhalent siphon: An opening formed by the mantle margins through which filtered water, waste, sperm and glochidia are expelled; located dorsal to the inhalent siphon.
Foot: A large muscular extension of the body that is used for digging, locomotion and, in juveniles, feeding.
Furrow: A shallow depression on the shell; sometimes referred to as a sulcus.
Gills: Thin paired structures in the mantle cavity that are used for respiration, filtering food and, in females, brooding young.
Glochidium (plural – glochidia): The larval form of a freshwater mussel that attaches as an external parasite to a vertebrate host, usually a fish, where it metamorphoses (transforms) into a free-living juvenile mussel.
Gravid: A term used to describe a female mussel that is brooding her young.
Growth ring: A dark ring on the surface of the shell that indicates a period of rest following an interval of growth.
Hermaphroditic: Able to produce both sperm and eggs; may be capable of self-fertilization.
Hinge: The portion of the dorsal margin of the shell where the two valves are held together by an elastic ligament.
Hinge teeth: Tooth-like structures along the dorsal margin of the shell that help hold the two valves together (pseudocardinal and lateral teeth).
Inflated: Swollen or expanded.
Inhalent siphon: An opening formed by the mantle margins through which water, food and sperm are brought into the body; located ventral to the exhalent siphon.
Interdentum: A flattened area of the shell between the pseudocardinal and lateral teeth.
Iridescent: Exhibiting rainbow colours.
Labial palps: A pair of structures on either side of the mouth that separate food particles for ingestion from non-edible particles.
Lateral teeth: Elongated teeth that extend along the hinge line of the shell.

Glossary (continued)

Left valve: The left valve is on your left when looking down at the top or dorsal part of the shell with the anterior end pointed away from you.

Ligament: A tough, elastic-like material that connects the two valves of the shell at the hinge.

Mantle: A sheath of tissue inside the shell that encloses the body of the mussel, secretes the shell material and serves a sensory function.

Marsupium (plural – marsupia): A pouch in the female gill that contains the developing embryos.

Muscle scars: Areas where the anterior and posterior adductor and retractor muscles attach to the inside of the shell; usually very distinct.

Nacre: The inside layer of the shell; often iridescent and referred to as mother-of-pearl.

Periostracum: The thin, fibrous material covering the outside of the shell.

Posterior ridge: A ridge on the back half of the outside of the shell extending from the beak to the posterior ventral edge.

Posterior slope: The area on the dorsal part of the shell between the posterior ridges.

Pseudocardinal teeth: Triangular-shaped hinge teeth located near the anterior end of the shell in front of the lateral teeth.

Pseudofeces: Particulate material filtered from the water but released instead of being ingested.

Rays: Coloured lines, usually green, that radiate from or near the beak toward the ventral margin of the shell.

Right valve: The right valve is on your right when looking down at the top or dorsal part of the shell with the anterior end pointed away from you.

Serrated: Notched or grooved like the blades of a saw.

Striated: Marked with a series of raised lines or ridges.

Truncated: Having an end shortened or squared off.

Ventral: The bottom edge of the shell, opposite the beak and hinge.

Wing: A projection of the valve that extends dorsally above the hinge line.

Glossary taken in part from Cummings and Mayer (1992), Nedeau *et al.* (2000) and Parmalee and Bogan (1998), with the permission of the authors.